I0454538

Lessons From A Martian Invasion

Copyright © 2023 Wayne McRoy

All Rights Reserved

ISBN: 9798867825096

You're listening to the Alchemical Tech Revolution and I am your host Wayne McRoy.

Good evening, everyone tonight we're going to look at some lessons from a martian invasion and this is actually primarily taken from a work from a gentleman named Hadley

1:49

Cantril, who worked for the Princeton radio project under the auspices and direction of a Mr.

Paul Lazarsfeld.

And this is kind of a continuation of some of the other lessons

2:05

we've been going through here recently relating to social engineering.

So this was a large portion of some of the early beginning phases of how they used the broadcast media for purposes of social engineering.

2:21

And this is one of the studies, one of the early studies where they determined that this medium of radio was a very important facet for the social engineering of society, they had discovered through the use of various methods, that people could be easily influenced by radio.

2:47

So they went ahead and they did a large research study into this.

And part of that research study was actually the presentation on October 30th 1938's War of the Worlds

Broadcast, now made famous by Mr.

3:02

Orson Welles,

over the CBS broadcasting system,

the Columbia Broadcasting System. Before it actually got involved in television,

it was a largely involved in radio.

Okay, so this was one of the big stories of the day back then.

3:19

So what was done was this was actually an experiment and many people don't realize that this was an experiment in social engineering, and this gentleman, Hadley Cantril, in 1940, wrote a

book based just on this one experiment, this one radio experiment and that's what we're going to look at tonight.

3:40

We're going to read in Chapter 3 of his book,

talking about some of the findings of this radio broadcast of War of the Worlds, and how it was done, and how it affected people on a psychological level, and how this could be used later for different types of methods and research.

4:00

So we're going to get right into it here.

No other

broadcast has produced a panic comparable to the one which found several million American families, all over the country, gathered around their radios, listening to reports of an Invasion From Mars.

These reports were brought to them over a national network from New York City,

4:19

our greatest metropolis where people should know what is going on. Both the form and the content of the broadcast seemed authentic. As one listener put it,

I just naturally thought it was real,

why shouldn't I? Even this program

did not affect more than a small minority of the listeners.

4:37

If we are to explain the reaction then we must answer two basic questions.

Why did this broadcast to frighten

some people when other fantastic broadcasts did not? And why did this broadcast frighten some people

but not others? An answer to the first question must be sought in the characteristics of this particular program, which aroused false standards of judgment in so many listeners, and I'm going to pause there for a second.

5:04

So this is saying that fasel standards of judgment in so many listeners,

so this is the suspension of disbelief,

essentially is what this is, this term has been come to be known, right?

So back in 1938 when this broadcast occurred, many people did not have this whole suspension of disbelief notion down pat yet.

5:28

Because radio at that point was largely used to transmit information.

This is where People mostly got news stories at the time and we'll get a little bit more into that as we read on here as far as why and how people bought into this whole fictional narrative.

5:45

Well, let's read on. Realism of the program in spite of Dorothy Thompson's remark that nothing whatever about the dramatization was in the least credible,

No matter at what point The Listener might have tuned in,

No one reading the script can deny that the broadcast was so realistic for the first few minutes,

6:03

That it was almost credible to even relatively sophisticated and well-informed listeners. Miss Thompson accepted the sheer dramatic excellence of the broadcast must not be overlooked.

This unusual realism of the performance may be attributed to the fact that the early parts of the broadcast fell within the existing standards of judgment of the listeners.

6:27

And I'm going to pause there.

Pay attention to those words.

Existing standards of judgment now, this is a very important

Idea. This falls into what you would call normalcy bias.

Okay?

So people had this normalcy bias, this is what they would expect when they would turn on the radio on any given night of the week or I think this took place on a Sunday night.

6:47

If I remember correctly on a Sunday night, they turned on the radio and they expected certain things on there to happen and most of the broadcast, the beginning phases of it lined up with these things that they expected to happen.

So this is what they call here existing standards of judgment.

7:04

So that being the case, you can see how they have this normalcy

bias.

They'll turn this on.

This is what they expect to hear.

It's all in their routine.

This is what they expect.

Like, if you turn on the television on a particular night of the week and you know there's a certain show that's supposed to be on,

7:20

Imagine that show was on like normal and everything seemed fine and then there was a break in the middle of the program, where they made some announcement and it looked and sounded official.

It's the same kind of thing that went on here.

But let's read on here. By a standard of judgment, we mean an organized mental context which provides an individual with a basis for interpretation, if a stimulus fits into the area of interpretation covered by a standard of judgment, and does not contradict it, then it is likely to be believed.

7:53

And I'm going to pause there.

And this is still a standard used today, a standard of judgment, right?

So if you present the information to people in a way they're familiar with, and they're in a format that they're expecting something particular here and you present it in that way, then they're more likely to believe it.

8:15

If they think it's actually something out of the ordinary and it's breaking into the normal pattern of things, so to say.

So let's go ahead and keep reading on, just what some of the more accepted and common standards were that provided interpretations for the immediate acceptance of the broadcast as news are given below. Later

8:38

in our discussion, we shall be concerned with the problem of discovering more individualized standards of judgment when it counted for the persistence of the original interpretation, even though the events reported became quite fantastic.

So we're going to pause there again.

8:53

So it's important to realize there's individualized standards of judgments that have to be considered for things as well.

But we'll get there.

Let's read on and see

what else is said in this Princeton radio research study. Radio as accepted vehicle for important announcements, the first wide use of radio in the country was to broadcast election returns.

9:20

Since that time, important announcements of local, national, and international significance have been repeatedly made. A few short weeks before this broadcast, millions of listeners had kept their radios tuned for the latest

news from a Europe

9:35

apparently about to go to war.

Remember folks, this was 1938, think of the context of the time.

So now what this is saying here is, radio had been known to be the medium for getting breaking news from around the world and flocally, and stuff like that, and they would sometimes occasionally break into the regularly scheduled programs to give you this news.

10:01

So, that being said, that's kind of the context we're looking at here.

So let's read on.

They had learned to expect that musical programs,

dramas, broadcasts of all kinds would be cut off in a serious emergency to inform or warn an eager and anxious public, a

10:23

large proportion of listeners,

particularly those in the lower income and educational brackets have grown to rely more on.

the radio than on the newspapers for their news.

The confidence people have in radio as a source of news is shown in the answer to a question asked by the magazine, Fortune, poll which of the two radio or newspaper gives you news freer from prejudice.

10:49

And this remember this was in 1938 this took place.

And this poll that Fortune Magazine did was probably in the same time frame. 1939 maybe in 1940, right?

So it says here, 17 percent answered newspaper. 50 percent, believed, radio news, was freer from prejudice.

11:09

Well, the rest either thought both media were the same, or didn't know which was, which one was less prejudiced on this particular night.

When the listener tuned to the Mercury Theater, he heard the music of Ramon Raquello and his orchestra coming from the Meridian Room in the Park Plaza Hotel of New York City. Soon

11:32

after the first piece had begun, an announcer

broke in, ladies and gentlemen, we interrupt our program of dance music to bring you a special bulletin from the Intercontinental radio news, with our present distance, it is easy to be suspicious of Intercontinental news, but in the context of the program, such skepticism is reduced.

11:53

This report brought the story of the first explosions on Mars.

The music was resumed only to be followed by another break.

Ladies and gentlemen, following on the news given in our bulletin, a moment ago the government meteorological borough has requested the large observatories of the country to keep an astronomical

12:11

watch, this bulletin contains the information that a huge flaming object believed to be a meteorite fell on a farm in the neighborhood of Grover's Mill,

New Jersey. The swing band gets in 20 seconds more, then the invasion continues uninterruptedly.

12:29

Almost all of the listeners who had been frightened and who were interviewed mentioned somewhere during the course of their retrospection

is that the confidence they had in radio and their expectation that it would be used for such important announcements.

A few of their comments indicate their attitudes and I quote, we have so much faith in broadcasting in a crisis.

12:50

It has to reach all people.

That's what radio is here for, and quote, the announcer would not say it if it was not true, the announcer would not say it.

If it was not true, they always quote, if something is a play and quote, I always feel that the commentators bring the best possible news.

13:12

Even after this, I still will believe what I hear on the radio and quote, it didn't sound like a play, the way it interrupted the music when it started, end quote.

So let's continue on reading here.

Those were some quotes of people interviewed in this Princeton radio experiment. Prestige of speakers, it is a well-known fact to the social psychologist, the Advertiser, and the propagandist, that an idea or a product has a better chance of being accepted, if it can be endorsed by, or if it emanates from some well-known person, whose character, ability, or status is highly valued, the effect of this prestige

13:56

and it is especially great when an individual himself has no standard of judgment by means of which he can interpret or give meaning to a particular situation that confronts him,

And when he needs, or is interested in making a judgement or finding a meaning, the strange events reported by the announcers in this broadcast were so far removed from ordinary experience and yet it gave such great potential and personal significance to the listener that he was both bewildered and in need of some standard

14:26

of judgment, as in many situations where events and ideas are so complicated or far removed from one's own immediate everyday experience that only the expert can really understand them, here, the layman was forced to rely on the expert for his interpretation.

14:44

And I'm going to pause there for a moment,

folks, let's read that again as in many situations where events and ideas are so complicated, or far removed from one's own immediate everyday experience,

that only the expert can really understand them, here, the layman was forced to rely on the expert for his interpretation.

15:06

Where have we seen that before, right?

Where have we seen that before?

Whereas something just belabors any type of common sense, you know, for a narrative that you must depend upon an expert to tell you what to do in this case, right?

15:25

And this is as old as time itself, folks.

These are some of the control mechanisms used by the social Engineers of this world and they've been time tested and true to be successful, right?

And this is just something, this is laid out.

15:42

This is a publication from 1940 studying the 1938 War of the Worlds broadcast on CBS radio, right?

So they knew what they were doing.

It was an experiment, they wanted to see what kind of reactions they could get from the public.

15:59

And I think they got all the data that they needed to understand ways in which they could deceive the public and make them believe

something that is false is true, right?

So, that's essentially what's been done here, but let's continue reading

16:16

and see what else that Mr.

Cantril has to say here. The Logical expert in this instance was the astronomer, those mentioned, all fictitious

agents. Were Professor Farrell of the Mount Jennings Observatory of Chicago,

Professor Pierson of the Princeton Observatory, Professor Morse of McMillan University in Toronto, Professor Indel Coffer of the California Astronomical Society and astronomers and scientific bodies in England,

16:47

France, and Germany, Professor Richard Pearson, which was Orson Welles, was the chief character in the drama. When it called for organized defense and action,

the expert was once more brought in, General Montgomery Smith, commander of the state militia at Trenton, Mr.

17:06

Harry McDonald,

Vice president of the Red Cross ,Captain Lansing of the signal Corps.,

and finally, the Secretary of the Interior described the situation, gave orders for evacuation and attack, or urged every man to do his duty.

17:22

It is interesting to notice that only the Office

of the Secretary of the Interior was named here.

The Listener was affected entirely by the institutional role and status of an unnamed speaker.

The institutional Prestige of the other experts and authorities is obviously more meaningful and important than the individuals themselves.

17:45

And I'm going to pause for a moment there.

So they really pulled the wool over people's eyes.

Didn't they?

So, you know, that being the case like, Think about the context of this, okay?

This news report comes across the radio, people accept it as a true thing.

18:05

They start naming off all these experts in their giving names, even though they're fictitious names of these supposed experts in various places that, you know, reinforces the whole idea that this must be true and people who obviously are right in that time and place there, they're kind of in this

18:26

fight, or flight type response, which is exactly where they want

you folks, they want you to be reactive, right?

They want you to react, they don't want you to think

critically.

So that's exactly what this did.

This reinforced, the idea of this reactive state, this fight or flight response in people.

18:43

This fear response, right?

People actually took this seriously and responded to it in a serious way.

So that being the case, I mean, they did

so, without actually taking any time to really research any of this stuff or look deeper, dig deeper into what had been done here.

19:06

So you know we could see how malleable the public mind can be in a situation that they perceive as an emergency, right?

They won't necessarily take the time to really dig deeper and try to find the truth of what's being presented to them.

19:23

They react instinctively,

It's a reactive state.

So they put you in that reactive state, and you're less likely to try to really dig into the details of things if you think you are in imminent danger, which is exactly what they did with this radio broadcast, but let's continue on and see what else

19:41

Mr. Hadley Cantril has to say here in this work.

This dramatic technique had its effect. Quote:

I believed the broadcast as soon as I heard the professor from Princeton and the officials in Washington, end quote. Quote: I knew it was an awfully dangerous situation

20:01

when all those military men were there and the Secretary of State spoke, end quote. Quote: if so many of those astronomers saw the explosions, they must have been real, they ought to know, end quote. Gonna pause there.

Let's read that one again because I find this one amusing, because this is still the same argument people use,

20:22

Well everybody would have to be in on it. Listen to this:

If so many of those astronomers saw the explosions, they must have been real,

they ought to know.

That's what

one of the listeners said, right?

So, here we go.

Here is the same kind of argument.

20:39

Well, do you know how many people would have to be in on it?

Well, only the fake ones that the phony names that they give you in the news media that you never bothered to read.

Research to see if they're really even actual people or if they ever had that opinion or said anything, right?

And that's essentially what's been done here.

20:54

So you can see this is a blueprint, right?

A social engineering blueprint for how to use broadcast media to steer and control the minds of people to get what they want from them.

Okay?

And this was one of the early pioneering studies in this to see just how much people would buy into this stuff.

21:13

So that being the case.

Let's go ahead and we'll read on here, specific incidents understood - The realistic nature of the broadcast was further enhanced by descriptions of particular occurrences that a listener could readily

21:28

Imagine. Liberal use was made of the colloquial expression to be expected on such an occasion.

The gas was a sort of yellowish green.

The cop warned, one side there, keep back.

I tell you, avoid shouts, the darn things, unscrewing, an example of the specificity of detail in is the moment of Brigadier General Montgomery Smith, quote: I have been requested by the governor of New Jersey to place the counties of Mercer and Middlesex as far west as Princeton and East to Jamesburg under martial law.

22:03

No one will be permitted to enter this area except by special pass issued by state or military authorities.

Four companies of state militia are proceeding from Trenton to Grover's Mill, and will aid in the evacuation of homes within the range of military operations, end quote. Particularly frightening to listeners in the New Jersey and Manhattan areas were the mentions of places well-known to them.

22:26

The towns of Grover's Mill, Princeton, and Trenton, New Jersey were featured early in the broadcast. Plainsboro,

Allentown, Morristown, the mountains by the Hudson, Hutchinson River Parkway, Newark, the Palisades, Times Square, Fifth Avenue, the Pulaski Highway, the Holland Tunnel are all familiar to Jersey and New Yorkers.

22:50

And listeners throughout the country

would certainly recognize many of those names as real. Quote: When he said

Ladies and gentlemen, do not use route number 23, that made me sure,

end quote. Quote: I was most inclined to believe the broadcast when they mentioned places like South Street and the Pulaski Highway, end quote. I'm going to pause for a second there.

23:13

So you see when they use actual real details

from the real world to present these narratives

in certain ways, it makes them more believable.

And, you know, that being said, if people are familiar with certain places or familiar with the names of these places, they would recognize them as a real place and think, well, this must really be something that's really happening, right?

23:38

And that's it.

Essentially, one of the things here that was taken advantage of by this radio study.

Okay.

They wanted to see if they could pull this off.

Essentially, and they did and let's read on and see what else is being said here.

23:55

Now, there's another person here that they're quoting from the study, quote: if they had mentioned any other places but streets right around here, I would not have been so ready to believe, end quote.

So get this.

So the guy says here that because they were street names that he knows right in the area, that that's what made him believe, right?

24:16

Even though, you could probably look outside and see these very same streets and not see anything.

Not there, but he didn't bother to do that, did he?

That just shows just how, you know, malleable the human mind is to these type of suggestions, right?

As long as you're hitting them, with this whole suspension of disbelief, kind of a motif here or what do they call it here?

24:38

A standard of judgment, right?

That that's what the researcher in this paper calls

It, standards of judgment, as long as it falls within their normal or expected standards of judgment,

They're more apt to believe it.

And that's exactly what this does.

This takes advantage of that type of a principle.

24:55

So let's read on here. Everybody believed

the events reported proceeded from the relatively credible to the highly incredible.

The first announcements were more or less believable,

although unusual to be sure.

First, there is an atmospheric disturbance, then explosions of incandescent gas, a scientist then reports that his seismograph has registered a shock of earthquake intensity.

25:20

This is followed by the discovery of a meteorite that has splintered nearby trees in its fall.

So far, so good.

But as the less credible bits of the story begin to enter, the clever dramatist also indicates that he too has difficulty in believing when he sees when we learn that, the object is no meteorite.

25:40

But a metal casing, we are also told that the whole picture is, quote: a strange scene like something out of a modern Arabian nights, end quote.

Fantastic.

That the more daring souls are venturing near before we are informed that the end of the casing is beginning to unscrew, we experience the announcer's own astonishment, he says here, quote: ladies and gentlemen, this is terrific.

26:06

When the top is off, he says, this is the most terrifying thing I have ever witnessed.

This is the most extraordinary experience.

I can't find words, end quote.

A few minutes later, Professor Pierson says, quote: I can give you no authoritative information for their origin or their purpose here on Earth.

26:25

It's all too evident that these creatures have scientific knowledge far in advance of our own,

It is my guess, after the battle at Grover's Mill between the thing and the soldiers in the Jersey farmlands tonight, they are the vanguard of an invading army from the planet

26:42

Mars, end quote.

So I'm going to pause there for a minute, so you can see back in the time, over the broadcast medium of radio this would have a very profound effect on

26:59

People, think about this, many people thought this was something that was really happening simply because of the way the material was presented, right?

So the material was presented to kind of infiltrate their accepted standard of judgment of the time.

27:20

So they tuned in expecting to hear just music from the hotel up there in New York, right?

From Raymond Raquello and his orchestra or whatever

it was claimed to be at the time.

That's what they tuned in to hear.

That's what they thought was going to be on.

27:36

And they were just, they were already primed at this point for listening to breaking news stories on the radio because of events happening in Europe at that time and that, the fact that war may have broken out

any moment over in Europe.

So they were listening intently to hear any kind of breaking news for that thing, and they knew that there might be periodic interruptions in the regularly scheduled programming for news events like this.

28:01

And that's when they presented this War of the Worlds scenario.

Never told anybody it's a story, right?

This is fiction, a work of fiction.

It's a dramatization.

They never announced that for people.

So, most of the people that tuned into this, they weren't expecting this.

28:20

They didn't know that's what it was and they accepted this as real news because they presented it

like they would real news stories that people were familiar with at the time.

So This falls into that whole idea of what their judgement of the time would be, their normalcy bias,

28:37

so to say, that's what we call it nowadays.

So anyway, let's read on though and see what else

that Mr.

Cantril reveals in this report. The bewilderment of the listener is shared by the eyewitness when the scientist is himself

28:55

puzzled, the layman recognizes the extraordinary intelligence of the strange creatures, no explanation of the event can be provided. The resignation and hopelessness of the Secretary of the Interior counseling us to place our faith in God provides no effective guide for action, no standards of judgment can be

29:15

applied to judge the rapid fire of events.

Panic is inescapable.

I'm going to pause there for a minute.

So by using people's standards of judgment against them, in this case, and then presenting them with this fantastical idea that we were being invaded by martians right there, their giant machines,

29:38

and, you know, they were spreading gas and using all kinds of weapons against people and laying waste to two cities.

This was out of the commonsensical view of the world at that time, right?

29:54

People did not accept that kind of thing as being possible but here it was being presented to them as if it was a true factual thing.

So it kind of hit on their suspension of disbelief and they didn't know how to react.

And so it says here, panic is inescapable and that's exactly what they were trying to accomplish by using this radio broadcast in that way.

30:15

Seeing how much Panic they could stir up.

And once again, panic being the word of the day, Pan – ic, based upon the archetype of Pan.

And I won't get into that tonight, but that's a whole different story altogether, but it's hitting up on a key

30:35

archetypal type of an energy inherent in the world and in mankind and setting up people for things to come, but let's read on.

The total experience, careful observation of everyday life, behavior ,or careful

30:52

introspection of one's own reactions in the course of an ordinary day

indicate that in social life, the normal individual experiences patterns or configurations of social stimuli. It is the atmosphere or the effect of a social situation that we notice long before

31:09

we are able, if we happen to try to analyze precisely what it is in the situation that creates a particular characteristic, in pressing us. The football fan wedged in between enthusiastic alumni, listening to the bands and the shouting, watching the teams, has the experience of being at the football game, an experience which is to be sure composed of the various stimuli impinging upon him, but an experience which results from the perception of all these stimuli

31:39

as patterned as coming together, as being inextricably interwoven in the production of a stimulating experience that

he may have traveled miles to experience. And I'm going to pause there.

And this is what it's all about.

It's the entire experience, okay?

31:57

It's the entire system of experience.

And this is what the science of cybernetics is set up to understand, cybernetics being the science of systems control.

And when you look at social behavior as a system, then you could understand what a lot of this is all about and how cybernetics methodologies have been used to

32:21

Steer the behaviors and patterns of society since at least this time, and we're talking back in 1938, probably before that as well, it wasn't referred to in the same terminology as we use now.

32:36

But the standards and the methods are all essentially the same.

It works the same, no matter what you call it, the methods work, the same. Let's read on. A person

in church is likewise experiencing a social situation with particular characteristics, that he can describe with adjectives meaningful to him even the awe or deference one may feel in an empty Cathedral seems to be more of an immediate perception than an accretion due to a series of related specific past experiences.

33:09

The importance of creating the proper atmosphere, conducive to any desired action is of course well known to the revivalist, the Cardinal, the dramatist, and especially today, the dictator, the elaborate preparations made by Hitler and Goebbels for their National party.

33:28

Celebrations are recognized musts for them.

If they are to enlist the enthusiasm they want to demonstrate,

It is obviously the total effect

they are after, just as a composer

keeps his whole theme in mind while writing separate bars of a symphony.

33:44

The lights, banners, uniform,

Airplanes, marching, singing,

and speaking at Nuremberg Congress has all gone to make up the experience of a parka tog and to reinforce adoration of der fuhrer.

And I'm going to pause there for a moment.

34:00

Boy, it sounds like this guy seems to have really idolized Hitler, doesn't it?

Hmm.

I wonder if there's something to that with a lot of these socialist types and social engineering types.

Hmm.

So you see kind of the mind state of the people behind this study, don't you?

34:23

But remember, this was before the war had started.

Well, actually, this book was published during the war.

Let's put it that way.

But at the time that the original radio broadcast here, the War of the Worlds, radio broadcast, the war hadn't really started just yet or hadn't really begun in earnest just yet, so let's read on.

34:49

In our discussion, we have broken the program down into what we regarded as important characteristics engendering belief.

This type of analysis could easily be extended further by showing how individuals have been conditioned to more specific items in the drama.

35:04

But the extension of this method puts a false emphasis on the problem by assuming, at once, that a social stimulus is essentially a series of discrete elements to which people have somehow

learned to react to. The enormously important possibility, which our approach so far has overlooked, is that social stimulus situations have their own characteristic and unique qualities.

35:28

These qualities in here, in the total pattern or configuration of the stimulus, just as the characteristics of it triangularity, or circularity in here in certain figures.

And I'm going to pause there folks, and notice he uses the terms triangularity and circularity.

35:46

And these are both essential terminologies to cybernetics methodologies.

The cybernetics approach to things because you see, circularity is one of the main components of what you could call feedback or a feedback loop which is an important facet to cybernetically engineering a system of some sort into a certain configuration.

36:13

Alright and don't get the

term cybernetics confused because we've been taught to conflate that idea with robotics and computer systems and artificial intelligence and things like that, which are certainly a part of it.

But cybernetics, in and of itself is merely defined as this: cybernetics is the study of whole systems control, so it's the study of the control of whole systems.

36:38

So it looks to take an entire system and learn the most effective ways to control the system through various means.

Looking at the big picture.

That's what cybernetics is.

It's looking at the big picture, how do we control and steer the big picture, not so much focused on a microcosm of the whole, but focused on the whole itself.

37:02

So that's why some of these ideas are inherent in a lot of these things and this actually predates what would later come to be known as the term for cybernetics because it's right around this time frame that cybernetics as a science was pretty much introduced and brought into the world roughly in this time period.

37:24

The mid to late 1940s

and onward.

So, you know, this, this was the early phases.

Many of these research projects here were some of the things that brought the cybernetics group into fruition and the cybernetics methodologies into fruition.

37:43

So let's read on here.

I don't want to get too

up on those different points because we got a bit more to go.

This broadcast of Martian Invasion, certainly had an atmosphere or structure all its own and the methodological device we have necessarily employed of describing one thing at a time should never obscure the fact that we are dealing with a situation experienced as a unit.

38:09

And I'm going to pause there folks.

Essentially, he just said what cybernetics is all about, right?

Let's read on. For some persons,

certain specific elements may have been more important in the total experience than others.

The case studies show enormous variety but no experience reported seems meaningful if entirely isolated from the whole context, the elementary from which the element of terror springs.

38:37

Inevitably from the method of the investigation, not from the experience of the subject.

If anyone doubts this, let him reread the "As reported" at the beginning of the second chapter, which by the way, in the second chapter of this book, it gives a lot more individual statements of people who heard the broadcast and how they reacted to it and stuff like that.

38:59

So that's what it's talking about.

So it's saying here that it's the method of Investigation, not the experience of the subject that really matters here within the context.

So I don't agree with that, I think that's double talk.

39:16

I think he knows exactly that

it's all about the reaction of the subject.

That's what they were looking for.

That's what they wanted.

That's what they were studying.

So, you know, it's not so much about as he says here that the method of Investigation, it's about the reaction of the subject.

39:35

So it's the total opposite of what he's saying here.

But anyway, let's read on and see what else it says here. Tuning in late in spite of the realism of the broadcast, it seemed highly unlikely that any listener would take it seriously.

Had he heard the announcements that were clearly made at the beginning of the hour?

39:53

He might then have been excited even frightened, but it would be an excitement based on the dramatic realism of the program, there would not be the intense feeling of personal involvement.

He would know that the events were happening out there in the studio, not right here, in his own state or his own country.

40:13

In one instance, a correct

valid detached or dramatic standard of judgment would be used by the listener to interpret events. In another instance, a false realistic or news standard of judgment would be employed.

So let me rephrase that gobbledygook right there.

40:32

So, basically what he's saying here is, once again, it goes back to the idea of what's called a standard of judgment or we would call it a normalcy bias nowadays.

So if you hadn't heard that brief little blurb that they played before the program began,

40:48

just prior to the program beginning, you wouldn't have known that

this was a fictional representation or a radio play.

You would have just tuned in with your regular normalcy bias and expected that you'd be listening to the music being played down at the hotel, right?

41:08

And that's what people were expecting when they tuned in.

So if they didn't tune in just a few minutes prior to the beginning of the broadcast, they wouldn't have known this was going to be a dramatic representation or a fictional story.

So when they tuned in, they expected to hear the band music playing.

41:25

And when it was interrupted by this news report, well, they thought it to be a legitimate news report because weeks prior to that they've been conditioned into accepting the normalcy of having breaking news reports interrupt the regularly scheduled program because of events happening in Europe.

41:43

So, their minds had been conditioned to accept this, and this isn't by accident folks, this is not coincidence.

These kind of things do not happen by accident or by coincidence.

So just to understand what's being said here.

42:01

So what he's saying here is, had the person heard the announcement prior to the start of the show, he would have used a different standard of judgment to react to it.

He would have realized it was a story and maybe enjoyed the story.

But the standard of judgment most people listening, or a good portion of the people listening, was that they thought this was real news being told, and just for standards of judgment today, whenever you turn on the news on the television, understand that you're using the wrong standard of judgment if you think that's real news.

42:40

So it's the same game being played today, okay?

They take advantage of this concept of the standard of judgment of people.

So people will tune in expecting that this is an authoritative source and that what they're saying is true, right?

42:57

And not expect that they're pushing fiction on you ,which much of the newscasts are fictionalized.

Okay, they may take, maybe, some little snippets of true things and work them together in a story, but then they'll show you of a clip

that's six years old,

that's supposedly Ukraine right now.

43:14

And then, they'll pull on your heartstrings by telling you some sob story about what happened to some children over there and much of that is probably false, what they're presenting on the television. Not saying nothing's happening there, I'm sure there are many things going on there right now.

43:34

And you know, I can't imagine it's good but what they're presenting on the television is not the truth of the matter.

Let's put it that way.

So it's the same story today.

They use this this concept to manipulate people's minds and their behaviors and their responses and the public reaction to things, its mass psychology,

44:00

and that's the important part, even though the individual is smart, the crowd is stupid.

And this is well known in mass psychology, and that's why they cater to the crowd.

So anyway, let's continue on here.

We got a bit more to get through.

The number of listeners, who dialed to the program

44:18

after the preliminary announcement, may be approximated by information obtained in two separate investigations. The data from each of these studies, furthermore, amply demonstrate that the time a person tuned in was a major determinant in shaping his later reactions. In a special survey conducted for the Columbia Broadcasting System,

44:37

CBS, the week after the broadcast, interviews were made throughout the country on 920 persons who had listened to the broadcast. Among other questions asked, were, at what part of the program

did you tune in?

And did you realize it was a play?

44:55

Or did you think it was a real news

Broadcast? Forty-two percent said they had tuned in late.

And as table one shows, there is a very pronounced tendency for those who tuned in late to accept the broadcast as news.

And for those who tuned in at the beginning, to take it as a play, only 12% of the person's interviewed listened from the beginning and thought they were hearing a news report.

45:18

So I'm going to pause there.

So they made the little announcement beforehand right before the show started, that it was a fictional representation, but a lot of the people didn't catch that, for whatever reason they tuned in late or didn't catch it or anything else of that matter.

45:34

So they didn't know that it was fiction.

So, of course, they took it as a news report and this all comes down to trust in the media, right?

And this was the early days of broadcast media.

So people at this point really didn't have much reason to distrust it because, by and large, it was giving them accurate information in the beginning phases.

45:58

And this study was one of the early studies that showed that if they wanted to take things a whole separate way, they could shape the reality with this broadcast media,

to make it represent things that it may not necessarily represent on its own.

46:14

And that's exactly what they've done.

It's a science called social engineering.

And they've mastered it and through the use of mass telecommunications, they've made social engineering so much easier for themselves.

This would not be possible, were it not for broadcast media.

46:32

It would be much more difficult to control the way society behaves without things like radio or television, or the internet or computers, or social media, or anything of the sort that we have now.

Look at how they've tailored it down through the years and made it down to the point where they could custom social engineer individuals.

46:55

That's exactly where we're at right now, they could socially engineer you on an individual level.

Whereas, back in those days, it was kind of hit or miss.

Like, they wanted to hit the whole crowd, so they had to use mass psychology strategies to do this.

47:12

Whereas now they can tailor make it down to the exact individual, simply because of all the data they've collected on you now.

And that's the big thing, data is king. We're living in the age of big data and it's soon to be, you know, well, we won't go into that tonight but there's big things coming.

47:37

The use of the data has made things possible

now that weren't possible back in the early days of these studies, but these are the foundational

aspects of this whole science of social engineering.

People need to understand, this is where it came from, this is how it began, these were the observations they made.

47:55

And this is what they based everything upon, and they discovered that the more data points they had, the more easily they could steer and control people.

So that's why they were always about collecting data.

And it's been a very quiet thing going on for many, many, many decades now.

48:11

But now, it's out in the open and it's become so easy with computer

algorithms to collect people's data, metadata, and everything else that they have real time.

Real time networks where they could actually try to manipulate your behavior and they've been very successful at this.

48:31

So, you know, that being the case, let's read on here and see what else it says in this study. In the survey made by the American Institute of Public Opinion, the question was asked, did you listen from the beginning?

Did you tune in after the program had begun? 61 percent answered that they tuned in after the program had started. 35 percent listened from the beginning.

48:54

Four percent did not remember as table 2 shows here.

Again, we find that those who tuned in late tended much more than others.

Let me skip past the tables here in regard to the broadcast as news.

Only four percent of the sample tuned in from the beginning and believed the broadcast to be a news report.

49:14

So, I give credit to the those people at that time, they had half a brain.

Had they heard it from the beginning and understood that this was a fictional representation,

I find it kind of interesting that there were still four percent of people who heard it from the beginning and knew what was being presented was fiction, but they still believed it to be legitimate news.

49:35

So ,doesn't that say something about some of the people? What did George Carlin say?

I can't remember the exact quote, but it's something like imagine half of the people you know, are so stupid,

49:57

and then the other half is dumber than them, or something like that.

I know I butchered that quote, I can't think of it, I always found George Carlin funny, but, and not to get hung up on that, because I totally butchered that quote.

So, that's not even going to be funny now.

Anyway, but let's read on. Both of these studies led to the same conclusion that at tuning in late was a very essential condition for the arousal of a false standard of judgment to be sure many people recognize the broadcast as a play, even though they tuned in late, just why this was done and by whom will be discussed in the next chapter, but for our present purposes, it is important to raise

50:37

an to answer the question of how anyone who tuned in

at the beginning could have mistaken the clearly introduced play for a news broadcast? That is a good

question, isn't it?

So, analysis of these cases, reveals two main reasons why such a misinterpretation arose in the first place.

50:57

Many people who tuned in to hear a play by the Mercury Theater thought that regular dramatic program had been interrupted to give special news

bulletins, the technique was not a new one after their experience with the radio reporting of the war crisis in October 1938 and it was a more

51:15

usual procedure to accept such news reports as irrelevant to the expected program.

Then, as an integral part of it, of the 54 persons in the CBS study who listened from the beginning and thought the broadcast was a news report, 33, which is 61 percent of them, said that the interruption seemed to them authentic.

51:36

This is apparent from the comments. Quote:

I have heard other programs interrupted in the same way for news broadcasts, end quote. Quote: I believed he was interrupting the program for a news flash, end quote. Quote: the news was presented in such an authentic manner, end quote. And isn't that the case?

51:56

So, you can see, the way that many people react to this kind of stimulus.

So we could see that even though they tuned in from the beginning, and they knew they were being presented with a fictional story, when they had interrupted the music and the various portions of the broadcast to bring these news reports,

52:23

these couple people 33 out of how many?

They say, 33 out of 54, believed it to be real news anyway, because of how it was presented. So, you know, what do you say about that?

52:43

What else could you say, gullible much?

Anyway, let's read on. The other

major reason for the misunderstanding is the widespread habit of not paying attention to the first announcements of a program.

And I'm going to pause there.

52:58

This is actually an important point.

Many people do not pay attention to the commercials or to the beginning points of a broadcast.

When you, or somebody, when you recognize the show hasn't started yet, and you hear jibber-jabber, blah, blah, blah, blah, blah, you're not really paying attention.

53:15

It's the same on the television.

If there's something else on there, it's not the show you're looking for, you're not really paying attention, it's just background noise, right?

So this is an important idea and they understand this and they know this and that's why they put the announcement in a very short blurb in the beginning of the program, it was all part of the experiment, folks.

53:36

Let's read on. Some People do not listen attentively to their radios

until they are aware

that something of particular interest is being broadcast.

Since the beginning of the hour is concerned with station identifications and often with advertising, it is probably disregarded by about 10% of the 54

53:54

people who misinterpreted the broadcast, although they heard it from the beginning, said they had paid no attention to the announcements.

These people, obviously just happened to be tuned to the Columbia Station and were not like the others who erred anticipating The Mercury Theater.

54:10

My radio had been tuned to this station several hours.

I heard loud, talking and excitement and became interested.

My radio was tuned to the station, but I wasn't paying attention to it.

We had company at home and we were playing cards while the radio was turned on.

I heard a news commentator interrupt the program, but at first did not pay much attention to him.

54:31

I started to listen only, when the farmer began giving a description of the landing of the tube.

So you see all these people here, just, I'm going to pause here.

You see these people here, they had the radio on, it was background noise, right?

54:46

It's this kind of the same as a lot of people with their television set.

Today, we leave it turned on, and it's largely background noise.

It's ambient noise.

You're not really paying attention to it, but you know it's there and maybe once in a while something loud will come on and might catch your attention for a minute, and that's kind of what happened with some of these people.

55:05

That's one of the important facets of this whole study.

They understood that.

This was likely to be the reaction here.

Let's read on.

Anyone who studies the characteristics of radio knows that one of its chief

55:21

shortcomings is its inflexibility as far as time is concerned.

The listener must be at his dial at the right moment

if he is to hear the program. In this respect, print obviously enjoys an enormous advantage. Newspapers, magazines, and books can be read when it is convenient to read them whereas a radio program exists for a few brief minutes and then disappears forever.

55:46

Remember, folks, this was written in 1940.

Anyway, it was kind of a novelty at the time, but it really caught on as a communication methodology for people and a lot of people really liked radio.

56:04

It was a very popular thing and it caught on.

And, you know, it was one of the standard mediums of communication for the day.

They got much of their news and entertainment from radio programs.

So you know, that being the case, it really caught on in a big way and this was one of the early stages of mass broadcasting.

56:26

That's why these studies were done to see how it could be used to affect people.

And sure enough, they had come up with different methods for affecting people and they put these experiments to work and they did

so in a large way with this War of the Worlds broadcast in 1938, and this is the result here.

56:48

This is the report.

Cantril really put together an entire book about this.

And we're only reading just a small fraction of the book.

And this book is available, it's out there on the internet archive, and you could find it for free online in various other places.

57:07

So, it's an interesting read.

If you wanted to pick it up to understand a little better, how some of these things have come about in modern society, and this would be the roots of the use of broadcast tools to manipulate the masses.

They knew what they were doing

57:23

when they put these experiments together, they were just looking for a way to fine-tune the data.

So to figure out exactly who's listening and where and when and why, and how, and all of those different little fine points to this whole thing, to better figure out how they can reach a certain category of people or a certain metric of people,

57:49

that's what had been done here.

Let's read on. The broadcaster can point out,

however, that comparatively few people do much reading. This disadvantage of radio has many practical consequences for the advertiser, the politician, or the educator.

58:07

The advertiser does not want to send his expensive commercial announcement into an air thinned of potential customers.

The clever politician

does not want to waste his best oratory before he has attracted the greatest possible audience. The late Huey P Long,

58:25

well aware of the radio habits of his constituents, began one of his radio talks as follows: Friends, this is Huey P Long speaking.

I have some important revelations to make but before I can make them, I want you to go to the phone and call up five of your friends and tell them to listen in.

58:42

I'll just be talking along here for four or five minutes without saying anything

special.

So you go to the phone and tell your friends that Huey Long is on the air.

What a great idea, man.

I'll tell you what, that guy,

one of the early pioneers of radio, it seems he knew what to do.

59:00

He knew how to get people's attention.

He figured he would tell people a little bit of nothing and then have them gather more people to listen before he wasted his breath, so to say.

Anyway, let's read on.

59:17

The great bulk of the latecomers consist of people who either turn their dials casually

at the beginning of the hour, trying to find something that pleases them, or of people who intended to listen to a specific program when it began,

but misjudged the time. The CBS survey showed that two-thirds of those who had tuned in late, did not know what program they wanted to hear

59:37

as they turn their dials.

Well, 12% of the latecomers had actually intended to listen to the Orson Welles broadcast at the beginning, too.

Being in late then is a normal aspect of the listening situation, but now we discover that tuning in late may lead to mass hysteria.

59:55

Such a phenomenon is so far rare, but might conceivably become important in times of crisis or national emergency, and I'm going to pause there.

Do you find his choice of words

very interesting there?

So he's saying here, tuning in late could cause mass hysteria, but such a phenomenon also, although so far is very rare,

1:00:17

but it might conceivably be important later in the time of crisis or national emergency.

So, you see how they've already introduced

the idea of using this as a social engineering trope

in this kind of a situation.

So let's read on and see what else he says here. In such situations,

1:00:35

it may be necessary to use different techniques to give news or information, perhaps wording a report in such a way that late listeners could understand it without becoming frightened.

This problem is important for our purposes.

Now, since we must discover, why approximately 50%, an unusually high proportion of the listeners to this broadcast, tuned in late as the combined figures of the American Institute and the CBS surveys reported above seemed to indicate, the large percentage of listeners who tuned in

1:01:08

on this special occasion, after the program had begun, seems chiefly due to two reasons in the first place, it must be remembered.

That the Mercury Theater program was competing with the most popular program of the week, that of The Versatile Wooden Hero,

1:01:24

Charlie McCarthy, the regular weekly survey of Hooper Incorporated, a commercial research organization, checking on the audiences of programs, estimated that the ratio of listeners to Orson Welles and Charlie McCarthy as 362,347.

1:01:41

According to restricted minute checks, the average family listens 48 minutes, out of this 60 minutes to the Charlie McCarthy program. Since McCarthy and his stooge Bergen were the recognized features of this competing broadcast,

it seemed probable that some people who did not listen throughout the whole hour would either turn off their radios

1:02:01

when the dummy act was finished, or would cruise around on the dial until they found something that interested them. If many persons did this, it is probable that they would misunderstand the nature of the Wells broadcast and keep their sets tuned to that program to learn more about the situation being so vividly described. I'm going to pause there, they knew exactly what they were doing folks.

1:02:24

It was not a misunderstanding of the nature of Wells broadcast.

They did it in the way that they did on purpose, they wanted to see if they could induce mass hysteria and they were successful.

Okay, these were some of the early aspects of psychology and sociology that they were looking at.

1:02:43

At this time, they wanted to know what it would do.

This this kind of behavior in people, so they knew full well, what they were up to here.

Okay.

So don't let this whole guise of plausible deniability in this report fool you about this, this guy knew perfectly

1:03:01

well what he was up to, he knew what he was going to be writing in this book.

He knew, and was just putting together the fine data points to try and reinforce the idea that they could use this tool,

this new tool of radio as a method for controlling people.

1:03:19

And they demonstrated it in spades here, and that's what had been done.

Anyway, let's read on here. To check this possibility, 846 cards were sent to persons all over the country known to have listened to the Mercury Theater broadcast.

1:03:39

They were asked if at any time during the hour, they had heard the Charlie McCarthy program, and if so, had they tuned out when Charlie McCarthy had finished.

His first act cards were returned by 518 persons. 18 percent reported that they had heard the competing program.

1:03:55

And 62% of these said they had tuned out when McCarthy had finished his first act and that they had, then kept their dials set to Orson Welles, the excitement of the Martian Invasion.

Then apparently stopped the dials of about 12% of Charlie McCarthy's devotees.

1:04:12

A second important reason for the increase in the number of rivals was the contagion the excitement created.

People who were frightened or disturbed by the news often hastened to telephone friends or relatives.

In the survey made by the American Institute of Public Opinion,

1:04:29

all people who tuned in late were asked, did someone suggest that you tuned in

after the program had begun? 21%

said yes.

So basically, this is all about herd mentality, right?

That's what this was about.

1:04:46

They wanted people to go out and tell everybody else.

Hey, you know what? Have you heard

what's going on?

Maybe you should tune in and hear what's going on, and this,

we see echoed decades later.

I remember exactly where I was on 9/11 and what I was doing and this was kind of the same thing, wasn't it?

1:05:06

It was the same thing.

If any of you are old enough, listening, I'm sure the vast majority of you listening are probably right around my age or maybe a little younger, maybe a little older, I'm not sure, but you probably remember very vividly that day on 9/11, what was going on.

1:05:23

I remember, I was actually working in route sales at the time and I was in between two of my accounts, driving and listening to the radio, when everything came on that was going on.

So, when I got to my next account, I went in, and I talked to the manager there and I said, hey, you hear what's going on in New York? and he's like, no.

1:05:43

I said, you got a TV?

He said, yeah.

So he brought out the TV.

So he turned it on and we stood there watching and we were just mesmerised by it, right?

And that's the same kind of thing that probably happened here with radio.

1:05:58

During this time, it's

through word of mouth, people began tuning in late, right?

So that being the case, you missed the beginning phases of it, and you wouldn't have known that

this was a staged event or a play.

1:06:16

So that's what happened back then in 1938 with this whole thing, and they knew, they knew, the aspects of mass psychology, to understand that this would potentially cause a panic and they just wanted to go ahead and catch the data points on that.

1:06:32

And we're going to stop it right there folks, because I think we got the point here.

So the lessons learned from a martian invasion, were methods in how to control the masses, by using different narratives,

1:06:51

and by using these different methodologies, trying to capture a hold of that idea of the normalcy bias and filter things through this normalcy bias to make it more believable to people.

1:07:07

So, even though they use different terminology for it in this particular paper here, this book actually, This is just a portion of the book, by Hadley Cantril published in 1940, on this whole event.

So, you know, even though they use the terminology a little differently, it's essentially the same thing.

1:07:27

So this would be affecting people's normalcy bias,

what we would call normalcy bias today.

1:07:46

So that being the case, this was the important facet of it.

They were trying to figure out what's the best way we could get people to believe that this is really going on, to believe this false thing.

1:08:02

Well, we have to go by what was their expectation.

What were they tuning in expecting to hear?

We have to keep that as normal and believable as possible and then maybe break in to maybe make them think in terms of,

1:08:20

well, this is an emergency situation now, and interrupts their normal routine and that they need to take it more seriously.

That's pretty much what they had done with this whole concept here.

So, there's all these different social engineering tropes that they use and this was a good illustration of it in many regards.

1:08:45

Even though the information presented would seem unbelievable, right?

Absolutely unbelievable.

They were talking about an invasion force from Mars.

That was the story

they went with here and people believed it.

1:09:01

They made people believe it by invoking several different techniques here.

One of the techniques is by actually using names and using the names of places that people were familiar with in the fictional narrative

to make people buy into the experience more, and that's what it's all about.

1:09:20

They created this experience in people's minds.

And they took advantage of the faith that the people of that time had in the radio broadcasting because they had seen it as an important arbiter of information.

1:09:36

And they trusted it, because up until that point, it hadn't been manipulated in this way to make people believe something that was false.

They were usually straightforward with people.

So that being the case, they utilized some of these different methods, and were able to steer people's judgment in a certain way.

1:09:59

So, you know, steering their judgment or what they called, their standard of judgment, what we would call normalcy bias.

Now using their standard of judgment as a weapon against themselves.

That's what they had done.

They had weaponized people's standards of judgment to make them believe the incredulous because they used their expectations against them.

1:10:23

So people's expectations tuning into this was,

well, I'm tuning into this entertainment broadcast and it's probably going to just be this music and that's it.

They tuned in thinking they were going to listen to the orchestra and this was interrupted a couple of times by what they thought was breaking news stories, which was a common practice at the time,

1:10:45

that they had gotten used to and they had been conditioned for up to that point, and of course, they thought it was legit because they had no reason to disbelieve it at that point, because they hadn't been lied to on a massive scale at that point

1:11:01

by this particular medium, like we have been lied to today on so many levels by all these different broadcast mediums.

So that being the case you know we could look back now with a different set of eyes on this whole thing and understand what it was, and we have these reports and stuff like this put out by these social engineers under the guise of trying to steer society into a good place or make humanity better or whatever.

1:11:32

Whatever excuses

they use to do these things that they do, right?

And they think they're doing good and promoting good things to fix the ills of society or whatever it is.

We have their information, their publications on this stuff now and we could look back and we could understand what's been done and we could understand why we're in the state we're in today.

1:11:57

And we can understand that at some point, regardless of what the original intentions were, this whole manipulative strategy has been abused.

It's been abused.

It's been twisted and perverted into something of a more controlling nature for a very small portion of society over the masses of society at large.

1:12:21

So that's what's been done here.

And now we could understand some of the methodologies and the lessons that they learned from some of these studies and it's important to look at these things and understand where this stuff comes from, how it's been shaped, how it's been financed, and a lot of this was financed by the Rockefeller Foundation.

1:12:41

Most of it, in fact. In fact, the whole Princeton Radio Project was funded wholly by the Rockefeller Foundation in the beginning, and it was later moved to, I think, Columbia University if I'm not mistaken.

It was no longer at Princeton after the beginning phases.

1:12:59

It was moved to Columbia University.

So that being the case, they studied many different aspects of radio and this was only one, just one of the experiments of that Princeton Radio Group at the time, this whole War of the Worlds broadcast.

1:13:16

So that being the case, we could see this is a lesson and example of how to socially engineer the masses into acting in a certain way that you want through

very subtle means, and like, this is just one of the basic methodologies used.

1:13:32

Now, imagine today with the technology that's available today, how much more they can steer an agenda like this or steer a program like this, an experiment like this?

Imagine, because of the level of data that they have at this point ,and the way they could steer it and control it down to the individual level,

1:13:55

Imagine what kind of manipulation they could pull off.

Now, when you look at this example,

It incited panic in a vast number of people and that was the intention.

They wanted to study the ramifications of it and they got some backlash for it in the beginning too, because people had a little more common sense back in 1938 when this occurred. Rght now, we've been socially engineered to have not as much common sense,

1:14:26

first of all, and also we've been engineered and conditioned to believe nonsensical things. I mean, look at our science.

Look at the things our science would have us accept. That 20 billion years ago, nothing exploded and became everything, and everything scattered around and then it all re-solidified again somehow, and then it rained and the rocks magically produced life and this new life found something to mate with and then boom, it became everything.

1:14:56

And it all happened by accident like some of the absurdities that they present to us as science.

And we could see that common sense went running out the back door when they presented these kind of ideas to us.

1:15:13

People back in the 1930s had a little bit more common sense.

They understood a little more about the natural world.

They weren't as far removed from the natural world as we are.

And therefore they were not indoctrinated as much as we are.

1:15:29

So, Voltaire said he who could make you believe absurdities can make you commit atrocities, and haven't we seen that?

Haven't we seen that since this time.

So you know many things have been done in the name of

1:15:46

absurdity.

Many atrocities have been committed in the name of absurdity and all we could do is look back now, and shake our heads, how did people fall for that?

Well, I think we just lived through something the past few years that we could understand the answers to how people could fall for something like that, right?

1:16:04

Especially when you look back at the events of World War Two and stuff like that.

Well, how did they, you know, fall for that?

Why did they go along with that?

Well, now we understand, because we've experienced something similar in our day and age, and still we question,

Why do most of the people go along with this, right?

1:16:23

It's the same kind of thing, and it all has to do with these social engineering strategies, that they gathered from many of these types of studies.

So that's why it's important to understand the foundations of where this stuff comes from and understand the methods that they used and these are some pretty basic methods.

1:16:42

Now, with the technologies they have today, they can do stuff in much more sophisticated ways than what they did back then.

But the methodologies are still the same.

The technologies have improved and been fine-tuned, but the methods are still the same.

1:16:58

So that's why it's important to understand that.

And that's why I wanted to talk about this tonight because, Invasion from Mars, Invasion from Russia, there's no difference, right?

It's the same methods.

Same strategies and and tropes being used, it's just a matter of making you believe certain things about who the enemy is.

1:17:21

And it's always almost always invariably a false enemy folks.

It's the boogeyman, the big boogeyman.

They always have to use this fear trope and keep you guessing and running around in fear and trying to maybe prevent something like this happening, or think that you have some kind of way of dealing with this.

1:17:46

Maybe if you wear a blue and yellow flag on your shirt, that that will help in some way.

Do you understand?

It's the same methods, regardless of what the specifics are about the situation.

1:18:05

That's always the same methods, and this is where they came from.

So that's why I wanted to touch on this, because as fantastic as the idea of an Invasion from Mars is today, it was even more fantastical in 1938 and yet they pulled it off, didn't they?

1:18:20

They got over a million people to believe it and fall for it,

hook, line, & sinker, to the point where it's said that the people in Grover's Mill,

New Jersey destroyed their water tower thinking it was a walker from one of these Martian Landings.

1:18:37

So anyway, at least that's the story I've heard.

So if that's true, the implications are staggering. Even if they could get a small percentage of the people to buy into something nonsensical like this and act upon it, it's still enough to do some serious sociological harm to the masses so that's why they do the things they do, and that's why it's important to look at this stuff.

1:19:03

So, I'm going to leave it there folks.

Thank you for tuning in tonight.

I hope this was educational and informative for people and I hope you had fun while you were here.

Thanks for tuning

In. We'll catch you next time.

Have a good night.

Visit www.alchemicaltechrevolution.com to further support my work. Thank you and God bless you all.

www.ingramcontent.com/pod-product-compliance
Lightning Source LLC
Chambersburg PA
CBHW062237290526
45794CB00006B/2324